D1532525

COWS

A PORTRAIT OF THE ANIMAL WORLD

DANYAL PAYNE

SMITHMARK

Copyright © 1997 by Todtri Productions Limited. All rights reserved.
No part of this publication may be reproduced, stored in a retrieval system
or transmitted in any form by any means electronic, mechanical, photocopying
or otherwise, without first obtaining written permission of the copyright owner.

This edition published in 1997 by SMITHMARK Publishers, a division of U. S. Media Holdings, Inc.,
115 West 18th Street, New York, NY 10011

SMITHMARK books are available for bulk purchase for sales promotion and premium use.
For details write or call the manager of special sales,
SMITHMARK Publishers,
115 West 18th Street, New York, NY 10011; (212) 532-6600.

This book was designed and produced by
Todtri Productions Limited
P.O. Box 572
New York, NY 10116-0572
Fax: (212) 279-1241

Printed and bound in Singapore

Library of Congress Catalog Card Number 97-066034
ISBN 0-7651-9215-2

Author: Danyal Payne

Publisher: Robert Tod
Editorial Director: Elizabeth Loonan
Book Designer: Mark Weinberg
Senior Editor: Cynthia Sternau
Project Editor: Ann Kirby
Photo Editor: Edward Douglas
Picture Researchers: Laura Wyss, Meiers Tambeau
Production Coordinator: Jay Weiser
Typesetting: Command-O Design

PHOTO CREDITS

Photographer/Page Number

E. R. Degginger 5 (top & bottom), 6, 12, 14–15,
16 (bottom), 17, 38, 39 (top), 54 (top), 59, 60,
64 (bottom), 65, 66, 68 (left), 70

Dembinsky Photo Associates
Mike Barlow 11 (top & bottom)
Dominique Braud 7
Adam Jones 34 (top)
Bill Lea 35
Sandra Nykerk 34 (bottom), 53 (right)
Stan Osolinski 7 (top), 13, 52–53
Carl R. Sams, II 40–41
Anup Shah 8–9
Bob Sisk 37–44
David F. Wisse 18–19

KK Productions 45

Picture Perfect 3, 24–25, 35, 39 (bottom),
46 (top), 48 (bottom), 51 (bottom)
Ken Akers 46 (bottom), 47, 48 (top)
Andris Apse 30 (bottom)
M. Barth 10
Robert Estall 4
Joe McDonald 42, 43
Nawrocki Stock Photo 64 (top)
Jose Raga 49

Tom Stack & Associates
Brian Parker 28
Bob Pool 36 (bottom)
Inga Spence 33 (bottom), 36 (top), 54 (bottom)

Lynn M. Stone 16 (top), 20, 21, 22 (top & bottom), 23, 26, 27 (top
& bottom), 29, 30, 31, 32, 33 (top), 50, 51 (top), 56–57, 58, 61 (top
& bottom), 62, 63, 67, 68–69, 71.

INTRODUCTION

The Lascaux Caves in France show evidence that prehistoric man interacted with the Bison that roamed the plains during the Paleolithic Age. Most likely, the wild animals were hunted for their meat and skins; cattle were not domesticated until much later.

There is something peaceful about being in the country, breathing in crisp air scented with the smell of grass and trees. As the feeling of peacefulness begins to settle in, various animals make their presence known: A blue jay may chirp boisterously, or a songbird sing a sweet melody. Yet there is a comforting sound that seems to be missing from the open countryside, one that can only be found on a farm or private land: the tranquil sound of cattle grazing quietly on lush fields of grass.

Cattle once roved the open plains in vast, wild herds. They can be found in ancient Roman and Celtic mythology and are immortalized in the night sky as the constellation Taurus, the bull. In many cultures, cattle are symbols of fertility and strength. In India, certain cattle are considered sacred and treated with reverence. In the United States and Argentina, Cowboys and Gauchos are mythic symbols of the indepent human spirit.

Today, cattle are one of the most valuable assets that mankind has. Cattle help us work; they feed us and clothe us. They provide us with milk and beef, and leather for clothing and necessary tools. And the work they can do in one day makes them priceless in many parts of the world. That's what cattle have become today: a basic necessity of life to millions of people.

ALL ABOUT CATTLE

Cattle belong to the family Bovidae. The family Bovidae can be divided into three groups: the first group having solid horns (antlers); the second, hollow horns; and the third, no horns at all. Cattle fall into the hollow horn category. They also have two to four teats, which are gathered at a single udder. Young are born one or two at a time, with their eyes open, and able to stand. They are related to other herd animals such as horses, antelope, sheep, swine, and goats.

Bovids are ruminants, which means that they chew a cud, tenderizing food to a pulp so it can be digested. Members of this family have multichambered stomachs and usually lack incisors. They are herbivores, living entirely on grass and thus requiring vast ranges of open land on which to graze. Although cattle can exist in the wild, the lack of open land for grazing restricts their numbers. But domestic cattle have been raised for food and draft purposes since the Stone Age, and continue to thrive in great numbers all over the world.

They are an integral part of many economies, from the least developed nations to the most industrial. Cattle futures are traded every day on stock exchanges, and politicians in rural areas battle daily over grazing rights. Historically, wealth of individuals and nations has often been measured in terms of the number of cattle they own.

In less than favorable conditions, labor cattle like these water buffalo enable farmers to save time, energy, and manpower, making them a very valuable asset in many parts of the world.

This oxen team demonstrates the primary use of certain breeds of cattle. Some breeds, like the ox, because of their sturdy builds, are better suited for labor than for dairy or beef purposes.

The thick, coarse coats of Highland cattle are best suited for harsh, cold climates, and are highly valued in areas like Scotland, northern Europe, and the northern United States.

Oxen teams continually prove that they can fulfill almost any aspect of heavy labor. Oxen are less expensive than machinery, and in the environments where oxen are used, they are usually more reliable.

Cattle are referred to by a number of terms. A grown male is a bull, a grown female, a cow. An infant is called a calf, and a yearling is an animal between one and two years old. A young female that has not given birth is called a heifer. Steers are male cattle that have been castrated and set aside for beef raising. Ranchers usually refer to their herds by the number of head. On breeding ranches, female cattle are called dams and male cattle are called sires. Oxen are labor animals, but the terms cattle are oxen are sometimes interchanged.

Cattle in the Wild

Though most cattle today are domesticated, many kinds of cattle exist in the wild. Shortages of open land and unchecked hunting has led to the endangerment and even extinction of many breeds. Vast herds of American bison, commonly called buffalo, once roamed the Great Plains of North America, where they were hunted by Native Americans for their flesh and hides.

With the arrival of European settlers, hunting buffalo became big business, and they were subjected to wholesale slaughter. At the

The North American bison is an endangered species. A few centuries ago, there were millions of bison, also known as buffalo, that roamed the North American plains. But due to unregulated hunting and habitat destruction, the bison came very close to extinction.

A small herd of musk-ox graze peacefully while their thick coats keep them warm in the cold northern climate. They get their name from the heavy, musky oil that they secret from their coats.

Following page: Wildebeest calves are an easy target for predators, and must stay close to their parents. Even within the safe bosom of the herd, they are still at risk.

beginning of the nineteenth century, over 60 million buffalo roamed the wilds of North America; by the middle of the century, they were extinct east of the Mississippi River. By 1900, only two wild herds remained on the continent. Protective laws passed later on helped increase their numbers, but only about two thousand remain today.

Many kinds of cattle, such as the Indian water buffalo and the yak, continue to live in the wild. However, even these breeds thrive mainly as domesticated animals.

Even though water buffalo are sturdy, hard-working draft animals, heat and strenuous labor can still take their toll. They need to rest occasion-ally out of the sun and near water.

These wildebeest are wild cattle found in various parts of Africa. Here, they make their way across the Seren-geti Plains as they migrate with the weather, seeking the best grazing land.

This Brown Swiss, a noted and prized dairy cow, is a sturdy animal which pro-duces excellent amounts of milk of especially high quality.

Domestication of Cattle

Cattle have been domesticated for about six thousand years. The wild auroch and the Celtic shorthorn—both oxen, believed to be the ancestors of most of today's cattle—were first domesticated during the Stone Age. Domestication began in what is now known as the Middle East and Africa; the notion was brought back to Europe and Asia by explorers. Early domestic cattle were valued primarily not for their milk or meat, but for their labor capabilities. But as time passed, more and more uses were found for cattle. Food, labor, and other uses of cattle have made them a valuable commodity.

Typically, farmers would choose the animal best suited for their particular climate and purposes. The sturdier and more resilient the animal, the more valuable it became to the farmer. Until selective breeding began to catch on in the eighteenth century, cattle were chosen based strictly on their natural characteristics.

There are two species of domesticated cattle today. *Bos indicus*, known as the zebu in India or Brahman in the United States, are cattle that have a large hump on their shoulders, right behind the neck. The other species, *Bos taurus*, includes most other breeds in the world. These two species are thought to share a common ancestor, *Bos primigenius*, known as the wild auroch, which was hunted to extinction in the mid-1600s.

Although cattle had been imported from Europe to the eastern United States for dairy, beef, labor, and other purposes much earlier, domestic cattle were first brought to the Great Plains when Spanish explorers tried cattle raising in the late seventeenth century. Cattle from the Old World had been in this hemisphere since Columbus landed at New Hispanola after his second voyage. They gradually moved north through Central America to the Great Plains, where they thrived on the vast stretches of unspoiled land.

In Uganda, this closely packed herd of Ankole cattle is probably making its way towards a holding place where they will be shipped off for labor, dairy, or beef purposes.

This zebu is searching for anything to graze on in these ruins in India. Cattle are sacred in India, meaning that they are usually left alone. However, this also means that many of them are left to their own devices without anyone to look after them or to provide care.

As oxen yield more than they consume in terms of labor and maintenance, their effect can be appreciated in very basic economic terms. They allow farmers to increase production at less cost than additional manpower or heavy machinery.

Holstein cows are one of the most popular dairy cows in the world today. They are sturdy, able to deal with weather conditions well, and reliable in the amount and quality of milk they produce.

As these water buffalo take a break, they receive a much needed washdown. Water buffalo must keep moist under the sun, otherwise their skin will dry and crack.

Breeding

In the cattle industry, breeding is the science of improving the characteristics of a type of cattle by controlling reproduction. It is a tricky science that takes a great deal of experience. In many cases, the art of breeding is passed on from generation to generation in a family. It is so specialized that often a breeder will concentrate on either bulls or cows within a particular breed; however, some do raise both sexes. The science of breeding is commercially motivated; the market is dictated by the demand, which in turn is determined by the individual needs of each farmer.

There are two types of breeding: straightbreeding and crossbreeding. Straightbreeding is the mating of cattle within a specific breed, and has two methods within it: outbreeding (mating cattle that are not related), and inbreeding (mating cattle that are related). Crossbreeding, which involves mating cattle from two different breeds, is a widely prac-

ticed method for improving a breed, but it is also an expensive and risky process with unpredictable results.

In modern times, technology has taken a major role in breeding cattle. Computers are used to keep track of milk and meat production as well as the growth rate of cows and bulls, which helps breeders make decisions in choosing the best breeding stock. Artificial insemination allows a bull of noteworthy quality to produce a great many more offspring than he would by natural means. This also allows breeders to import the frozen semen of a preferred bull directly to their farm instead of transporting the bull.

As well as controlling what animal is being bred, a breeder will control the number of calves born to a particular cow in her lifetime. This involves the fairly simple task of keeping the cows away from the bulls until a time that the breeder determines is best for

This bull, if possessing superior qualities, will be shown off at shows and fairs. He will also be used to breed: His semen will be sold and used to promote his characteristics in the same breed or help raise the quality of other breeds. He will continue to do this until a newer, "better" sire is found.

reproducing. A cow's gestation period is approximately 280 days. In general, summer is chosen for breeding so that calves will be born sometime in the spring, which is usually the best time for natural fodder. Births are usually uncomplicated, although problems can develop. A calf will take milk for about three months, after which it will be weaned and fed grain or other food.

Breed Societies and Herd Books

A breed society is an organization formed of breeders who are raising the same breed of cattle and who desire to improve and maintain the best aspects of the breed. There are rules for each society that are designed to protect the future of the breed. The breeders also work together to promote the various aspects of their breed for commercial sale.

A Herd book is the registry that keeps track of the animals registered in a particular society. These registries are sometimes published, and can be as extensive as registries for thoroughbred race horses. So that cattle breeders can trace the ancestry of a particular animal, Herd books include a wealth of information on each animal registered: any entered or registered name, date of birth, Herd book number, names and Herd book numbers of sires (bulls) and dams (cows), and sometimes a sketch showing any prominent markings. All of this information is given by a breeder after the birth of a calf. After a society has the information and it accepts the new calf, the calf is given a Herd number.

Some Herd books are closed, meaning that new candidates need to have their parents in the book to be registered, while other Herd books are open, and allow any animal that meets the society's requirements to be registered. There are also advanced registers that act as Herd books, but rather than recording the basics mentioned above, they record interesting, unusual, or outstanding achievements of a particular animal.

The Hereford is one of the top beef animals in the United States. Originally from England, the Hereford has been bred to meet the high demands of quality and volume. Since it has such excellent qualities, the Hereford has been used repeatedly for cross-breeding purposes.

19

The Brown Swiss cow is one of the top-rated dairy cows. Because of its natural qualities, it has not been cross-bred as much as other cows have been. A Brown Swiss, such as this cow found in Switzerland, can produce amazing amounts of milk and butter in one year.

On dairy farms, calves like this young Guernsey are fed both their mothers' milk and a prepared mixture designed to ensure growth and valuable qualities in adulthood.

thirty to forty gallons of water and produce eight gallons of milk. A cow is milked about twice a day and reaches her peak in milk production approximately two months after bearing a calf. After that, the quality of the milk usually goes down, but can be balanced with an adjusted diet.

After the cow chews and swallows her food, it goes into the first chamber of her stomach, called the *rumen*. If the food is chewed well enough, it goes onto the third and fourth stomach chambers, called the *omasum* and *abomasum*, respectively, where it is digested completely. But food that is not chewed well enough is either passed into the second chamber, called the *reticulum*, or brought back up for the cow to chew some more. The food that is brought back up for tenderizing is called the cud, and it comes from the first or second chamber of the stomach. After the food is completely digested, it moves into the bloodstream to nourish the cow and allow the udder to do its work.

The cow's udder is located between the hind legs and is supported by strong ligaments from the stomach and pelvis. A very important part of the cow, it is something that breeders keep a close eye on. A healthy udder should be attached with strong muscles and have evenly spaced teats that are neither too big nor too small, and that are pointed straight down. The udder comprises four glands that take elements from the blood and turn them into milk. The milk is moved from these glands to udder cisterns, located above each teat. The milk is drawn off from the end of the teat through a small opening controlled by tiny muscles.

Following page: These Holstein cows will produce many thousands of gallons of milk and hundreds of pounds of butter in a year. After a dairy cow has lived past its usefulness, it is usually shipped off and made into ground beef.

This Vermont farm benefits from a good amount of graze land, something essential for raising cattle. It would not economically viable to raise cattle if they could not graze naturally.

This Guernsey calf is only a few hours old. Dairy farmers used to place a calf near a cow that was not its mother to induce the flow of milk. However, using basic conditioning techniques, farmers have trained cows to produce milk even when there is no calf around.

Milking

A very important aspect of milk production is whether or not the cow will let the milk be drawn out. This is called milk flow. Essentially, the cow has to be in the mood to let the milk go free. There are various techniques involved in stimulating milk flow. Before modern techniques, a cow would allow milk to flow only if her calf was near. The calf did not have to be suckling; its presence was the only factor needed. If the calf died, however, other methods had to be used, such as rubbing another calf with the dead calf's urine or placing the pelt of the dead calf over another calf. Either of these methods would cause the mother to recognize the smell as that of her calf and let her milk flow.

Today, modern methods of stimulating milk flow have produced basic conditioned reactions in dairy cows. When they are faced with a stimulus associated with milking, such as being led into a milking parlor or having their udders massaged, dairy cows will allow their milk to flow easily.

Milking techniques

The various ways to milk a cow have changed with the passage of time. In the early days of dairy farming, cows would be milked by hand into a bucket. The milk would then be poured into a churn made of either wood or metal, then loaded onto a milk truck to be stored and processed. Later, farmers began milking straight into the churn, which was then

These hardy Guernsey, originally bred in France, make excellent dairy animals in terms of both the quality and the quantity of milk.

This Guernsey is ready to be milked. A cow will produce a significantly greater amount of milk when it is calving than when it is not, although cows have been conditioned to milk when they are not calving as well.

BEEF CATTLE

A great number of cattle are bred primarily for slaughter, and beef is a major industry throughout Europe, the United States, Canada, Argentina, and Australia. But the cattle industry produces a great deal more than meat. Cattle hide is used to make leather goods; bones are ground up and used in glue, fertilizer, and feed for livestock. Grease, gelatin, and margarine have cow fat in them, and some pharmaceuticals and dyes are made with cows' blood. Horns have been used as carry and signaling devices; manure is used as fertilizer and in some countries as heating fuel, after being hardened and dried out.

Beef and Veal

Meat, however, remains the primary goal of raising most cattle. Cattle produce two types of meat: beef and veal. Beef is the meat of a mature steer. Cattle are raised and fattened for about eighteen months, after which they are put onto the beef market. Cattle are ready for slaughter when they reach eleven hundred pounds. Veal is the meat from calves no older than twelve weeks. The best quality veal comes from calves that are milk-fed and then are killed when they are six weeks old. Veal calves are taken away from their mothers at birth and placed small pens that keep them completely immobile. There they are fattened until scheduled to be killed.

Europe is the largest beef producer in the world, with the United States, Canada, Argentina, and Australia as large producers as well. In Europe there is a much stronger

The Texas long-horn is one of the most recognized cattle today. Its impressive horns can reach over five feet from tip-to-tip.

The Red Angus cow finds its origins in Scotland. In the 1800s, they were introduced to the United States where they eventually became one of the most popular beef animals. Although the Red Angus is not a purebred, it is a very highly valued animal.

The Brahman (Bos Indicus) is one of two types of cattle found in the world today. In India, the animal is known as the zebu. The Brahman is primarily a beef animal and can be found all over the world in a variety of breeds.

The most exciting and dangerous event at a rodeo is the bull ride. Huge bulls, weighing up to a half ton, are ridden by cowboys for as long as they can stay on the agile, mean, bucking animals. Rodeo bulls are much less predictable than the horses that are ridden and are usually bred for their meanness. In the chute, the cowboy slips a heavily gloved hand into a handhold on a bull rope which is then tied tightly around the bull. The rope has a bell attached to annoy the bull, making it buck and spin that much more. While riding the bull bareback, the cowboy cannot touch the bull, the equipment, or himself with his free hand. He must stay on the bull's back for eight seconds; if thrown, he is totally exposed to the tormented bull's horns.

Brave rodeo riders will try and stay on a bucking bull for as long as possible using only one hand. Anxious rodeo clowns wait to distract the angry bull when the rider is thrown off.

Bullfighting is a very popular sport in Spain and Portugal. Fighting bulls are specially bred for this sport, possessing qualities of strength, speed, and a short temper.

Bullfighting

Rodeo is not the only professional sport devoted to cattle. Now the national sport of Spain, bullfighting was practiced in ancient Greece and Rome before it was introduced to the Spanish by the Moors during the eleventh century. A dramatic sport filled with pageantry, bullfighting is practiced in large outdoor arenas, and fights are preceded by lavish parades.

The object of the sport is for the bullfighter, called the *matador* (also *toreador*, or *torero*) to kill a wild bull, called a *toro*, with a sword. Fighting bulls are bred for the sport; they weigh over 1,000 pounds (370 kilograms) and are usually between 4 and 5 years old. In a typical bullfight, the bull is released and irritated by five toreros, who torment it with capes and lances. The role of the matador requires him to make elaborate, dramatic passes at the bull with a red cape until the dominated animal stands with its four feet square on the ground and its head hung low. The matador must then kill the bull by stabbing his sword between its shoulder blades and into its heart.

Although some people feel that bullfighting is an inhumane sport, others see it as an important part of Spanish cultural tradition. It is practiced in many Latin American countries as well, including Mexico, Peru, Columbia, and Ecuador, as well as in southern France. *Rejoneo*, a style of bullfighting practiced on horseback in which the bull is not killed, is popular in Portugal.

The bull run happens every year in Pomplona, Spain, to celebrate the feast of San Fermin. Bulls run freely through the streets, while the populace runs with them. It is an exhilarating and dangerous event that draws thrill-seeking tourists from all over the world.

CATTLE BREEDS

There are over forty different cattle breeds today, all of which have one thing in common: The cattle industry wants each breed to be the very best for its particular purpose. Some are bred for dairy, some for beef, some for both dairy and beef, and some for labor purposes. These purposes depend on such variables as topography and climate, availability of land, the economic status of a country, and environmental concerns.

In Europe, where land is not as spacious as it is in North and South America, there is a greater emphasis on producing cattle that include good dairy and beef characteristics in the same breed, while in the Americas, large herds of individual beef and dairy cattle can be raised. In less developed countries, cattle are used for all purposes, but mostly for labor and eventually for meat.

Breeders look for different qualities, depending on whether they are breeding for dairy or beef. Dairy cows must have healthy udders, while beef animals must be long and well-muscled, especially in the rump and thighs.

A unique animal because of its coloring, the Randall Blue Lineback is a rare breed used for dairy and beef purposes.

Although the Ayrshire's origin of strain is unknown, the breed today has been carefully bred and raised to produce a fine animal noted for its endurance and milk production.

These Friesian cattle, in one year, will produce tens of thousands of pounds of milk and hundreds pounds of butter. At maturity, they weigh close to a half of a ton.

African Buffalo

The largest of its kind, the African buffalo, or Cape buffalo, is considered one of the most dangerous animals in Africa. It can grow to over nine feet in length, with massive horns that are broad at the base and which curve slightly upward and inward. The color of its coat varies, but is usually dark. It usually feeds at night when it is cooler, and rests during the day. It has a mean disposition and has not been domesticated by man.

Angus cattle are the world's best producers of beef, in terms of both efficiency and quality.

Angus

George Grant first brought four Angus bulls from Scotland to the middle of the Kansas prairie in 1873. When the bulls first met the public, they were judged with a bit of skepticism. They were polled (naturally hornless) and a different color from that of the more popular breeds at the time.

Grant died five years later, and many of the settlers at his Victoria, Kansas colony later returned to their homeland. But the four bulls

The African water buffalo is considered to be one of the most tempermental animals in Africa. Its thick hide, immense bulk, and solid horns make it a formidable opponent to many predators, although the oxpecker perched between this bull's horns doesn't seem threatened.

left behind were to make a very distinct impression on the cattle industry. Prior to his death, Grant crossbred Texas Longhorns with the Angus. The resulting offspring proved to be sturdy and heavier than expected, giving promise as a valuable beef animal. Today, the breed is one of the most popular in the United States, where there are more than ten million registered Angus.

Beefalo

The beefalo results from breeding *Bison* (American buffalo) with any of the various breeds of domestic or exotic beef cattle. The best breed of beefalo was produced in 1960. Bison are often crossbred with other superior cattle to produce an animal that displays the best qualities of both species. Superior characteristics include better milking ability, better

The grass and feed that these Black Angus bulls consume will affect the quality of their beef as they mature. The bulls that are chosen for breeding will usually begin in the fall while others will continue to mature.

These Red Angus cattle will yield an excellent amount of meat when they are shipped off to slaughter. Beef cattle are measured in a variety of terms, including the amount of beef versus the amount of feed consumed, the quality of the beef, and the durability of the animal.

meat quality, and greater durability. Ultimately, the beefalo can be more cost efficient for a rancher. Of course, the characteristics of the resulting beefalo vary depending on the type of cattle chosen for crossbreeding with the bison.

Beefmaster

The Beefmaster is a deliberately developed breed that comes from Brahman, Hereford, and shorthorn cattle. Development of the breed began in the early 1900s and was perfected after 1930. While the Beefmaster was being developed, the cattle were kept in rough and strict conditions that would produce valuable characteristics in the final animal. The characteristics that were stressed by the rigid conditions now help the Beefmaster to perform well under normal and severe conditions on the range.

Ranchers look for the best performance in cattle. The stronger and more well-developed

an animal is, the more valuable it becomes. The ultimate achievement of the endeavor to breed only the animals with the highest endurance and fitness levels was the Beefmaster. This method of developing the beefmaster through generations of natural breeding with a focus on certain characteristics has implications for other breeds as well.

American Bison

These shaggy, heavy animals with bulky forequarters were once found all over the Great Plains and in other parts of the continent, their numbers ranging into the millions. Native Americans depended on the vast herds for food and clothing. But population expansion, hunting, and the Civil War took an extreme toll on the species. Conservation efforts helped save the American bison from extinction, but today only about two thousand bison remain.

Before the bison were hunted to near extinction, they were hunted by Native Americans who used the bison for food, clothing, building materials, and other miscellaneous uses.

Following page: Guernseys are highly valued animals because of their production of high-butterfat, high-protein milk, rich in betacarotene. They consume less than other breeds their size while still producing a superior product.

Australian Braford

The Australian braford is a crossbreed developed from Hereford and Brahman cattle from 1946 to 1952 in Queensland. The breed is characterized by the physical features of a Brahman, with the Hereford coat. It is a strong animal that works well in hot weather and even better in milder conditions. It can be found in Queensland and New South Wales.

Brahman

The Brahman breed in America comes from *Bos Indicus* of India, also known as the zebu. It is easily recognized by the hump on its shoulders and the loose, excess skin on its throat. The horns are relatively short and usually curve upwards. The coat is thick and may vary from red brown to black, with dark skin. The Brahman has a highly effective cooling system that involves efficient sweating to keep its internal body temperature down in hotter weather. The Brahman is a hardy breed that works well under adverse conditions, which makes it a popular beef breed in the United States, where they are raised and exported.

The Brahman was first introduced into the United States in the mid-1800s. But up until the 1930s, the breed was in flux, being continually experimented on through crossbreeding with animals imported from South America and India to finally produce the animal considered most satisfactory.

Brown Swiss

The Brown Swiss is one of the oldest and purest dairy breeds in the world. Raised in the mountains of Switzerland, it has developed characteristics of strength and endurance which are still evident in its descendants today. Imported from Switzerland in the late 1860s, the Brown Swiss then had one of the most tumultuous histories of any breed in the United States. It has gone on a roller-coaster ride of political maneuvering, disease, and transportation. But, through perseverance and the perception that the brown Swiss was not only one the best dairy cows but also purebred, milk farmers were finally able to bring the Brown Swiss into its own niche. Today, a Brown Swiss cow can produce in the high end of tens of thousands of pounds of milk and over one thousand pounds of butter in a year.

The Brahman was developed in the southern United States by interbreeding Indian cattle. The large, silvery-gray animal is especially resistant to heat and ticks, and is used primarily for crossbreeding purposes.

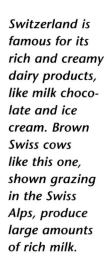

Switzerland is famous for its rich and creamy dairy products, like milk chocolate and ice cream. Brown Swiss cows like this one, shown grazing in the Swiss Alps, produce large amounts of rich milk.

Canadienne

The Canadienne has its origins in sixteenth-century France. As Canada was being explored and settled, this breed was imported from Normandy and Brittany. For a number of years after being imported, this breed didn't see very much crossing with other breeds. It was then later crossbred with other less hardy breeds in Canada to bring those breeds up to standard. It is primarily a dairy cow, well-suited to Canada's climate and natural foliage. Not a large animal, it has brown or darker colors, with occasional white splotches.

Charolais

The Charolais originated in France, getting its name from the region where it was raised. It was first introduced into Mexico after World War I. It didn't see introduction into the United States until the mid-1930s.

The Charolais was bred mainly for use as a labor animal and for producing meat in France, which has led to its popularity as a beef animal. Like the Brown Swiss, the Charolais suffered setbacks during importation because of disease, though by the 1960s, it was reintroduced as an import. It is a large breed with a light-colored coat, and can be horned or polled. It is a popular animal to crossbreed for raising beef characteristics of other breeds.

Chianina

Although thought to be one of the oldest breeds of cattle, the Chianina did not see introduction into the United States until the 1970s. It is a beef animal, having been raised in Italy for draft purposes. It has proved to be excellent in crossbreeding to improve beef quality in other breeds. The Chianina is colored white to gray, with well-developed muscles, and short, curved horns.

The Charolais was first bred in France centuries ago, and was prized for its ability to perform multiple tasks. A hardy draft animal, it also produced fine milk and beef. It has been bred for superiority in all three aspects.

Highland cattle are a tough breed coming from the mountains of Scotland. It seems to thrive on harsh conditions, including bad weather and poor land for grazing, and is able to survive where other cattle breeds would not. These qualities alone make it a popular breed to raise.

Originally from Switzerland, the Simmenthal is a highly valued animal for its dairy, beef, and labor characteristics. Today, the Simmenthal can be found in the United States, China, Australia, and Europe.

The Devon was first brought to the United States in the 1600s, where it was used as a dual-purpose animal. It is a tough breed, and works very well under adverse conditions and with natural foliage. Today, the Devon is raised mainly for breeding purposes but still retains decent dairy qualities.

Devon

This breed is another of the older breeds that exist today. Raised in southern England, this red-haired cow was originally bred as a dual purpose animal, with dairy and beef being the characteristics stressed. It was one of the first purebreds brought to the United States shortly after the first colony was established in the 1600s. The breed was developed in the United States mainly for beef, but it still retains decent dairy qualities. It has become useful to the beef industry because of its good body characteristics. It is also a hardy animal that works well under various conditions and with natural foliage. Its coat is bright red, and the animal can be either horned or polled.

Dexter

The Dexter is one of the smallest breeds around, not growing over four feet high at the shoulder. It was raised in Ireland and introduced to the United States sometime around the turn of the twentieth century. These small animals make good dairy and beef stock, giving more of both meat and milk in proportion to the food they need to survive. They do very well in hot or cold climates and work well with natural foliage. They have solid, dark coloring and short, curved horns.

Dutch Belted (Lakenvelder)

One of the more renowned and rare breeds is the Dutch belted. It is most notable for its unusual marking, a wide band that encompasses the torso of the animal like a big, wide belt. Dutch belteds have been used for crossbreeding with dairy and beef cattle, raising the quality of both.

The Dutch belted breed was not introduced to the United States until the late 1800s, when the Dutch, who were very protective of the breed, allowed it to be used in a circus exhibition. Today, the Dutch belted is a highly valued dairy animal.

The bull fight is a bloody and inhumane sport. A fighting bull is raised to maturity and then placed in the ring, where it will become enraged with spears and taunts from the matador until it is finally killed.

Herefords still command a strong place in today's beef market. Although the animal has had a tumultuous past, starting in England, it has become one of the more popular breeds because of its natural characteristics as well as its versatility in breed.

Fighting Bull

Known as *Toro de lidia* in Spain and *Brava* in Portugal, fighting bulls are bred for traditional bull fighting in Europe and many Latin American countries. Bred from ancient stock, these animals today are bred for one purpose only: to enter the bull ring. They have mean dispositions and physical agility. The color varies on these animals; their horns are of medium length and curve forward. They have long bodies with well-developed forequarters. The bulls are tested against each other before being put into the ring. In Spain, bulls are killed in the ring as part of the sport; in Portugal, they are spared in the ring and continue to fight until they have passed their prime.

Guernsey

The Guernsey finds its origins on an island off the coast of France, although today it is very different from its French ancestors. Through a great deal of artificial insemination, the Guernsey is now one of the best dairy cows in the United States. It has a high production of milk and butter as compared to the amount of food it takes in. The Guernsey is also an excellent range animal, working well with the climate and grass. Because of these qualities, the Guernsey is a popular dairy cow.

Hereford

The Hereford has a colorful history in America. Imported from England in the 1800s, this ideal beef animal has been the front-runner of beef cattle for a very long time. It was initially bred in England to provide the best beef qualities possible. The characteristics of the Hereford have allowed it to survive a demanding and changing beef market of the United States, making it still one of the premier beef animals available today. The Hereford is also used for cross-breeding purposes.

Holstein

The Holstein was bred in the Netherlands to produce an efficient dairy animal. It is usually black and white or red and white in color. They were first introduced into the United States in the 1800s. Today they are noted for their outstanding milk production, and are popular dairy cows all over the world.

The Holstein breed was first brought to America from Holland in the mid-1800s. The milk the Holsteins produced so impressed farmers that thousands of Holsteins were imported in ten years. Imports were stopped when cattle disease broke out in Europe.

Indian Buffalo

Also known as a water buffalo, this native of India and Sri Lanka lives in the jungles near water where it can wallow for hours or lie submerged with only the nostrils and horns showing. Like the Cape buffalo, it is a dangerous animal, sometimes attacking without provocation.

The milk from this animal is rich, yet its greenish color can put some people off from drinking it. The animal is also used for meat, and its hide makes strong, durable leather. Water buffalo are characterized by their thick, dark coats, and large, flat, ribbed horns that curve backwards. Water is an essential element for these buffalo, keeping their skin from drying out and cracking.

Jersey

The Jersey breed is an exemplary dairy cow, named for the island of Jersey in the English channel where it was developed. The mild, humid climate on this rocky island demands a unique method of dairy farming: Rather than grazing freely, Jersey cows are tied on a short chain and moved several times per day to make maximum use of the available grass. By the 1700s in the United Kingdom, Jerseys were already noted and bred for their superior milk, which is rich in butterfat, and legislation was passed to keep the breed pure. They were brought the the United States on English sailing ships throughout the 1800s, and they remain a popular dairy breed there as well. Jersey cows are distinguished by their wide foreheads and heavily-lashed "doe-like" eyes, which are large, brown, and clear.

Normande

The Normande is an excellent dual-purpose breed found in France, South America, and the United States. A sturdy animal, it adapts well to the various climates of these regions. It was raised in Normandy for thousands of years to reach the point where it now gives rich milk and good meat. They also work well for crossbreeding purposes.

Although the water buffalo found in Burma, India, and Thailand are domesticated and used for labor, the African water buffalo can be quite dangerous.

The Jersey is one of the best dairy cows found in the world today. A combination of durability, excellent dairy qualities, good economic returns, and international use has made this breed desirable since it was domesticated centuries ago.

Santa Gertrudis

This purebred animal was developed in the United States by crossing a shorthorn with a Brahman. The result was an animal that was as sturdy as a Brahman but yielded better meat with good milk supply. Its coloring is a deep red.

Shorthorn

A very versatile animal, this breed became very popular during the 1800s with farmers and settlers because of its ability to adapt, its usefulness as a labor animal, its good milk production, and good beef quality. The shorthorn is one of the oldest breeds in America and today is valued for its dairy and beef production, as well as its crossbreeding purposes.

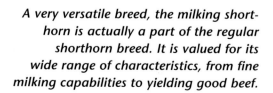

The Santa Gertrudis is a cross between a shorthorn and a Brahman. It is a sturdy animal with good beef and dairy qualities.

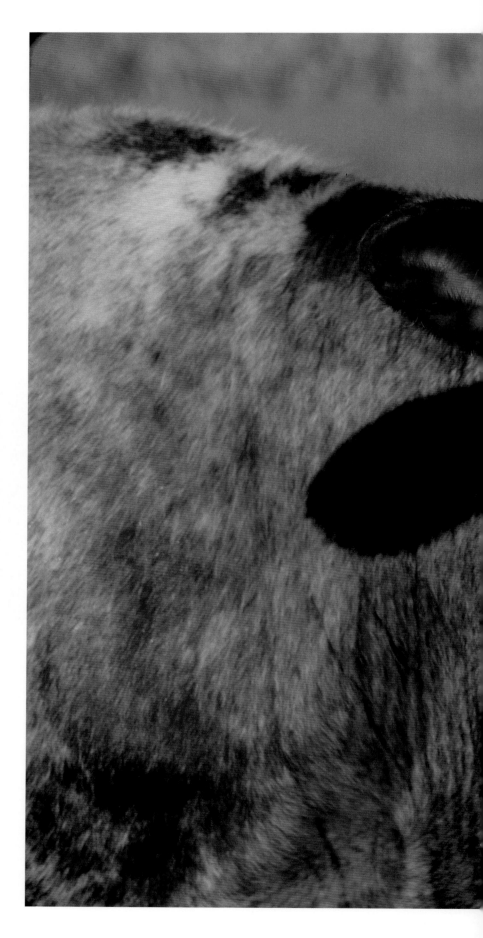

A very versatile breed, the milking shorthorn is actually a part of the regular shorthorn breed. It is valued for its wide range of characteristics, from fine milking capabilities to yielding good beef.

Texas Longhorn

Perhaps one of the most famous cattle known in America, the longhorn has seen a rough and demanding existence. It went from its introduction to the continent by the first explorers to its one-time status as the most popular and demanded breed. It then came close to extinction, only to return to its place today, where it exists relatively ignored by the cattle industry. It is distinguished by its lean body and by its huge horns, which curve up and outward, and often span over seven feet. Toughness and adaptability have been bred into the Longhorn for centuries, and these qualities may once again prove to make it a valuable breed in the future.

Yak

The yak has been domesticated in Tibet for centuries. Used primarily as a beast of burden, the yak is also used for dairy and other goods. The hide is used to make clothes, its dung is burned as fuel, and its milk is made into butter and an alcoholic drink called *tsamba*. A reliable food source and a sturdy draft animal, the yak has become a necessity of the Tibetan economy.

The yak is one of the most easily recognizable breeds of cattle. It is an essential part of life for people who have domesticated it, serving a number of roles, from draft animal to clothing provider.

During the height of the Cattle Kingdom in America, the Texas longhorn numbered in the millions. Years later, it came close to extinction but has been fortunate enough to have survived. Many cattle breeders think the longhorn may yet again have a future in the cattle industry.

INDEX

*Page numbers in **bold-face** type indicate photo captions.*